QUICK! Earth Science labs

Jennifer L. Boudart
Carol A. Thornton
Deborah C. Gutman

Illustrated and Designed by: Cox & Co.
Cover Illustrations by: Lopata Design

ISBN: 1-56911-807-8

Printed in China.

Table of Contents

Introduction

"Learning science is something students do. Full inquiry involves asking a simple question, completing an investigation, answering the question, and presenting the results to others."

National Science Education Standards

Quick Earth Science Labs provides active, hands-on science investigations in the content strands of earth science for students in grades K-2. Each unit in this book ties in with a specific program standard outlined in the National Science Education Standards. These connections are highlighted at the beginning of each unit. Investigations address fundamental earth science concepts including properties of earth materials, patterns in weather, and objects in the sky.

Stimulated by the challenge of simple questions, children:

- complete investigations or other organized activities such as structured games
- organize data or gather information to answer questions
- present results to others either orally or in writing
- apply what they learn to other areas

In the spirit of the National Science Education Standards, activities require students to use scientific inquiry to investigate a question, communicate findings, and connect learning to real-world experiences. Students are challenged to use scientific tools, and to work cooperatively when completing activities and sharing results.

Each unit investigation includes Teacher's Notes and these reproducible student pages:

Quick Lab Experiment—introductory activity

Data Collection/Communication—record sheet for Quick Lab findings

Working Together—cooperative group project

Using Tools—research using a scientific tool to gain information

Learning Link—interdisciplinary connection

You can reproduce student pages or use them as transparencies. Consider having students create a book to hold notes and drawings relating to the activities. Check the equipment table at the back of this book for materials needed in each unit as well. As children engage in the investigations, activities, and games, they will learn the essence of scientific inquiry—a refinement of youngsters' natural curiosity to learn and understand the world around them!

The Earth's Surface

The surface of the earth changes. Some changes are due to slow processes, such as erosion and weathering, and some changes are due to rapid processes, such as landslides, volcanic eruptions, and earthquakes... Fossils provide evidence about the plants and animals that lived long ago and the nature of the environment at that time. *National Science Education Standards*

Getting Started

To gather information online, enter the query: "planet earth" AND "children" AND "learning" or "earth" AND "classroom." Share background information below and books such as **Earth** (Andrew Charman, 1994) or **What's The Earth Made Of?** (Susan Mayes, 1995). Check the Equipment Table for materials needed to complete the activities in this unit.

About the Earth's Surface

- The earth has a diameter of 12,800 km (7,900 mi). It is believed to be about 4.6 billion years old.
- Earth is composed of air, water, soil, and rocks. It is arranged in layers: The crust is a relatively thin outer layer of the earth (about 40 km or 25 miles thick). The crust includes the land and the water we see, and is the layer we live on. The mantle lies beneath the crust and is Earth's thickest layer (about 2900 km or 1,800 miles thick). It is solid rock. The core is the innermost part of the earth. It is split into an outer core of liquid rock, and an inner core thought to be made of solid iron and nickel.
- The earth's surface changes. Rapid changes may result from events such as volcanoes and earthquakes. Slow changes may result from erosion and weathering.
- Fossils are plant and animal remains that became trapped in Earth's layers long ago. Fossils may form from bones or shells; or, they may be imprints of structures that have since disappeared.
- Maps and globes provide information about the surface of the earth: its water and land masses; landmarks such as volcanoes and mountains; and possibly climates, human-made boundaries and structures, and land use patterns.

Misconception

Children may believe the earth always looked the way it does today. In fact, Earth's land masses have drifted apart, life forms have evolved, and climates and landscapes have changed.

Teaching The Unit

Quick Lab Experiment and Data Collection/Communication: Provide enough caramel apples for each student (or pass out apple halves to start if you prefer). Begin the activity by discussing the earth and what might be inside it. Discuss properties of each Earth layer. Help students cut their apples in half and identify the following layers: caramel/nut layer = soil and rock; apple peel = crust (the thinnest and most explored layer); apple flesh = mantle (the thickest layer); core = outer core; and seeds = inner core. For Think About It, discuss how human activities and natural events change the appearance of the crust.

Working Together: This activity demonstrates how erosion reveals rocky mountains over time. Discuss how mountains might form on Earth, and whether or not they have always been here. Remind students that the earth is covered with rocks and soil. Mountains are made of rock. For the activity, students can collect rocks from the school grounds or from home. Be sure students spray their tubs for a long enough period of time, until the "mountains" begin to appear. You may wish to have students tilt their tubs to help soil runoff. Students should conclude that water carries away soil over time to reveal mountains, which is one way they "grow." This process takes thousands of years.

Using Tools: Discuss maps, globes and the way people use them. Review the vocabulary terms for the landmark labels. Help the class locate North America on the map and globe. The illustration in the activity is of South America. Gather state maps for children to observe in Think About It.

Learning Link: Discuss fossils and share books about fossils to begin the activity. Bring in clean chicken bones if possible, or substitute plastic bones from a skeleton kit. Help students choose additional objects to use for fossil impressions. For best results, use soft modeling clay or Play-Dough®.

Inside the Earth

PREDICT What does Earth look like inside?

LAB TIME

1. Look at the apple your teacher gives you.

   Think of your as a model of .

2. Slice your in half. Your teacher can help you.

3. at the picture of and what is inside it. has different layers inside. Read the name of each layer.

4. at your . Decide which part of the matches each layer of .

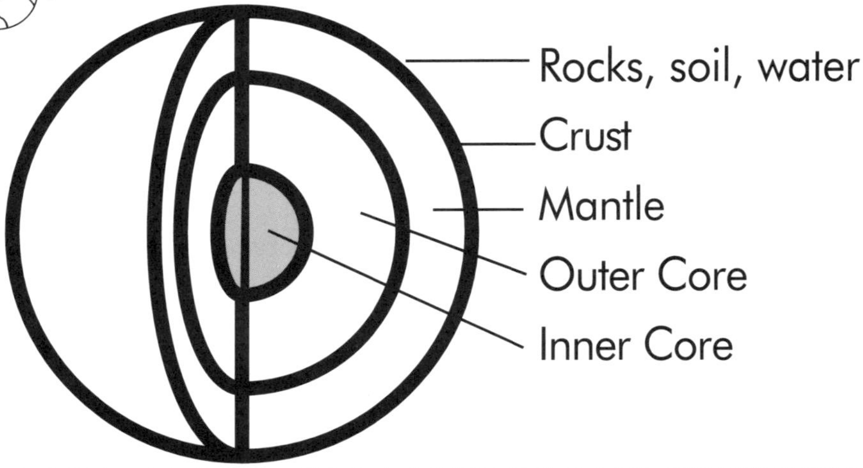

THINK ABOUT IT

Which layer is thickest? Which layer is thinnest?
Think of 3 words that tell about the inside of .
Tell or write about it.

Inside the Earth

Fill out this page as part of your lab.

LAB TIME

Decide which part of your  matches each layer of .

 the labels in the right places.

Rocks, Soil, Water	Inner Core	Mantle
Crust	Outer Core	

THINK ABOUT IT

How do people's actions change the crust of the  ?

Tell or  about it.

Mountains Rising

PREDICT

How do mountains "grow" from the crust of ?

LAB TIME

You'll Need:

3 big rocks | tub | soil | spray bottle | water

1. Place the in the . These are your .
2. Pour over the until you cannot see them.
3. Spray on the with the to make a rainstorm.

THINK ABOUT IT

What happens to the when you "rain" on it?

How does rain and washing over the land help grow?"

Do you think it takes a long time for to grow?

Tell or about it.

Map and Globe

Use a world map or a globe to learn more about the Earth.

A world map and a globe show where water, land, and landmarks like mountains are on Earth. People use these tools to find cities and other places.

LAB TIME

1. Look at the picture on this page.
2. Write the words in the right places on the picture.
3. Use a world map or a globe to find out what part of the Earth the picture shows. Your teacher can help you.

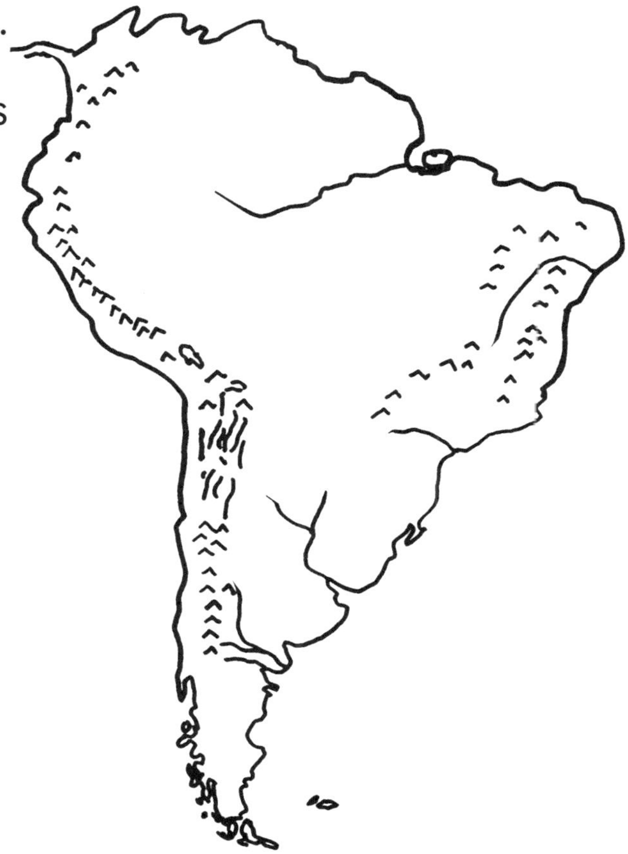

continent
island
ocean
river
mountain

THINK ABOUT IT

Look at a map for your state.
Find your city.
What landmarks are near where you live?
Draw or write about what you learn.

Learning Link

Fossil Match

The is very old. Over time, its crust has added layer on layer of and . Bits of once-living things are trapped in these layers. These bits are called fossils. Some fossils are rock-like parts of plants and animals. Other fossils are like footprints. They are shapes left behind by materials that have disappeared. Fossils can be millions of years old.

Make your own fossils and play a match-up game with a friend!

LAB TIME

You'll Need:

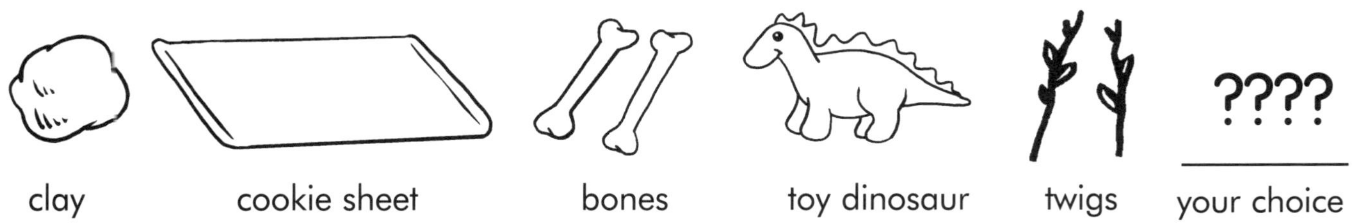

1. Cover the with .

2. Press your objects into the . Lift them out. at the shapes in the . These are your fossils!

3. Ask a friend to match each object to its fossil.

THINK ABOUT IT

Do you think the plants and animals that form fossils look like today's plants and animals?
Tell or about it.

Rocks and Soil

Earth materials are solid rocks and soils, water, and the gases of the atmosphere. The varied materials have different physical and chemical properties, which make them useful in different ways, for example, as building materials, as sources of fuel, or for growing the plants we use as food... Soils have properties of color and texture, capacity to retain water, and ability to support the growth of many kinds of plants, including those in our food supply. *National Science Education Standards*

Getting Started

To gather information online, enter the query: "rocks and minerals" AND "classroom" or "soil" AND "classroom." Also share background information below and books such as **The Living Earth** (Eleonore Schmid, 1994), **Under The Ground** (Pascale De Bourgoing, 1994), or **Let's Go Rock Collecting** (Roma Gans, 1997). Check the Equipment Table for materials needed to complete activities in this unit.

About Rocks and Soil

- The earth is made of rocks, soil, air, and water. All soil once was rock. Rocks don't become soil overnight. In fact, a layer of soil a few inches thick may take hundreds or even thousands of years to form. Rocks break down slowly due to weathering: the "attack" of wind, water, changing temperatures, and grinding against other hard surfaces. They become smaller and smaller, eventually releasing their minerals and turning into soil.
- The three rock types are sedimentary, igneous, and metamorphic. Sedimentary rocks form as layers of mud and silt settle to the bottom of rivers and streams and form soft rocks. Igneous rocks form from melted rock (magma) that seeps from inside the earth, then cools. Great heat and pressure change sedimentary and igneous rocks into metamorphic rocks, which are the hardest rocks.
- Rocks can be classified by the minerals they contain and the properties they demonstrate, such as hardness, color, weight, crystal formation, streak, luster, and cleavage.
- All soil is a mixture of inorganic (non-living) and organic (living or once-living) materials. Inorganic materials include rocks and minerals. Organic materials include plants and animals, and the remains of plants and animals that have died (called humus). Spaces between soil are filled with water and air.

Misconception

Children may not recognize that soil forms from rocks, or that new soil is always forming.

Teaching The Unit

Quick Lab Experiment and Data Collection/Communication: Before doing this activity, make screens for each student or group to use. Select wire mesh screens that are fine enough to catch larger objects but let dirt through. Cut the screens down to shoebox-size. Cover the edges with duct tape for safe handling. If possible, collect soil from school grounds, a wooded area, or other outdoor places. Be sure to scoop up leaf litter, rocks and anything else that lies in the soil. You may wish to create a class chart to summarize and compare students' findings with their soil samples. Discuss the fact that soil is a "living" material. Allow students to explore Think About It by looking at different soil samples if possible.

Working Together: Help students collect a variety of rocks or provide them yourself. Review the characteristics in the chart, pointing out the four sets of characteristics students will decide between for each rock. If necessary, practice identifying characteristics. Help students use their graphs to sort rocks according to each characteristic. For Think About It, allow time for students to pair up, or work as a class.

Using Tools: This activity demonstrates the ability of water to erode rock surfaces over time. Give each group a handful of small, rough rocks to work with. After shaking the can to model the action of running water, students will find bits of rock trapped in the coffee filter. Note: The action of rocks bumping against each other is also a contributing factor in this short-term experiment, and you might discuss the fact that in natural conditions, water carries rocks and knocks them against each other as well.

Learning Link: To make about 2 dozen mud piles, use one 12-oz bag of chocolate chips and 2 Tbs. of vegetable shortening, plus a few handfuls of the remaining ingredients. You can make one batch and use student volunteers. Or you can melt the chocolate and shortening, divide it and the remaining ingredients, and have students make their own batches. For Think About It, students may compare pretzels to sticks, nuts to rocks, gummy worms to real worms, and candy flowers to real flowers.

Quick Lab Experiment

Soil Search

PREDICT

What is  made of?
soil

LAB TIME

You'll Need:

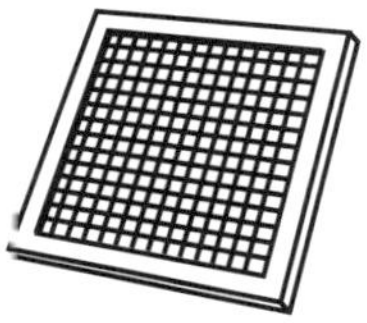
screen

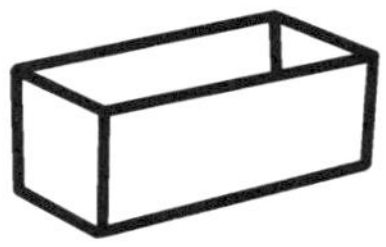
shoebox

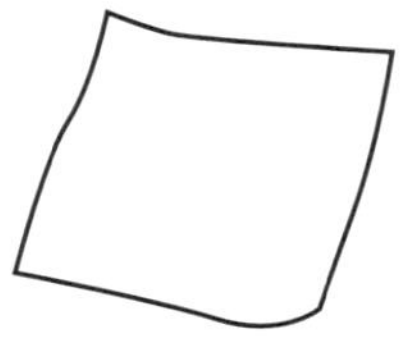
white paper

soil

1. Place a piece of  in the bottom of the 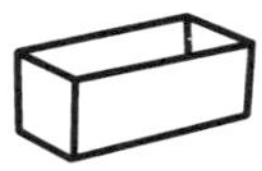 .
2. Place the 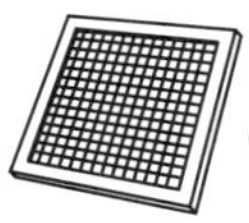 over the  .
3. Pour the  over the 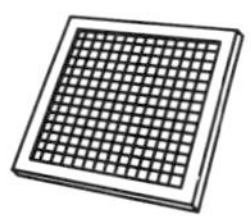 .
4.  at what lands in the 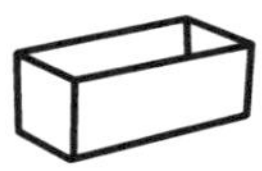 .
   Look
5.  at what gets caught in the 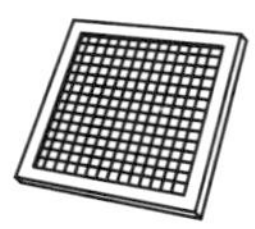 .

THINK ABOUT IT

Would you find the same things in  wherever you dig it up?
Why do you think so?
Tell or  about it.

Soil Search

Fill out this page as part of your lab.

LAB TIME

Draw or about what your has in it. Use the boxes.

plants	rocks
animals	other things ????

THINK ABOUT IT

3 sentences that tell about your :

1. ______________________________

2. ______________________________

3. ______________________________

Tell or about how and use .

Rock Sort

PREDICT

What are some ways you can sort into groups?

LAB TIME

1. Collect 5 . Your teacher will help you.

2. 1, 2, 3, 4, and 5 on pieces of tape. Stick a piece of on each .

3. at each . Fill out the chart. For each , an "X" in the boxes that tell how it looks and feels.

4. Use the chart to sort your . Use answers from one box at a time for each sort. Start by grouping all solid color .

	1 color	many colors	smooth	bumpy	mostly flat	mostly round	small	big
1								
2								
3								
4								
5								

THINK ABOUT IT

Read all the boxes you marked for one .
Have a friend guess which you are reading about.

Magnifying Glass

Use a magnifying glass to see how changes .

has the power to change ! As washes over it carries away tiny pieces from the surface. Over many thousands of years, water can carve canyons, shorelines, and turn big rocks into small rocks.

LAB TIME

Work with a partner.

You'll Need:

plastic can with lid · bowl · coffee filter · magnifying glass · small rocks · water · timer

1. Wash your . Place them in the .
2. Fill the almost halfway with .
3. Place the lid on your . Shake it for 15 minutes. Take turns shaking and watching the .
4. Hold the over the . Pour the and into the .
5. Wait 5 minutes for the to dry. Use the to look at what is trapped in the . What do you see?

THINK ABOUT IT

Where would you find being worn away by ?
Where do you think sand comes from? Tell or about it.

Mouth-Watering Mud

Make a mixture of mud you can eat! Your teacher can help you.

You'll Need:

bowl · spoon · chocolate pieces · vegetable shortening · freezer bag

waxed paper · pretzels · trail mix · gummy worm · candy flowers

1. Pour and into the . Melt this "mud" in a microwave for 1 minute, 30 seconds. Stir with a until the mud is smooth.

2. Place in the . Crunch the into small pieces with your hands.

3. Stir , and into the mud.

4. Use the to drop piles of mud onto the . Sprinkle on top of each mud pile.

5. Cool the mud in the refrigerator. Enjoy!

THINK ABOUT IT

How is your mud like real mud?
Tell or about it.

Weather

Weather changes from day to day and over the seasons. Weather can be described by measurable quantities, such as temperature, wind direction and speed, and precipitation. *National Science Education Standards*

Getting Started

To gather information online, enter the query: "weather" AND "children" AND "classroom." Also share background information below and books such as **How's The Weather: A Look at Weather and How It Changes** (Melvin Berger, Gilda Berger, 1993) or **Weather** (Seymour Simon, 1993). Check the Equipment Table for materials needed to complete the activities in this unit.

About Weather

- Weather is the result of short-term changes in the atmosphere, including: temperature (measure of warmth or coldness); precipitation (condensed water that falls from clouds as rain, hail, sleet, or snow); humidity (the amount of water vapor in the air); clouds (a large mass of moist air containing tiny suspended water droplets or ice particles); air pressure (the amount of force with which air "pushes" against a surface per square inch); wind (moving air that results from changes in air pressure and travels parallel to the ground); and the sun's heat.
- Long-term weather patterns create climates.
- Instruments that measure weather include: nephoscope (wind direction); anemometer (wind speed); hygrometer (humidity); thermometer (temperature); barometer (air pressure); and rain gauge (precipitation).
- Meteorologists study Earth's variations in temperature and precipitation that create different weather conditions and climates. They use this information to make predictions about the weather.

Misconception

Children may believe all regions of the world experience seasons at the same time of year and in the same way.

Teaching The Unit

Quick Lab Experiment and Data Collection/Communication: Make five copies of the Quick Lab page for each student, in order to provide a set of symbols to cut and paste for each day of the school week. Discuss different kinds of weather events, and relate them to the symbols. Explain that the day's record might include more than one symbol, e.g., if it is rainy, cloudy, and windy. Also explain that students should choose the dominant weather for the day. For Think About It, discuss seasons, seasonal weather in your area, and how people dress for and deal with weather.

Working Together: This is a whole-class activity that requires students to form four groups as indicated on the Working Together page. Organize students in groups and have them practice making each of their rainstorm sounds. Remind students that they must wait to make their sounds until they see their group number written on the chalkboard. They should continue to make their sounds until their number is erased. Read the following story aloud, posting and erasing group numbers at points indicated:
The sky is dark and filled with clouds. A rainstorm is on the way. First, soft rain begins to fall (write "1" on the board). Next, wind begins to blow (write "3" on the board). The rainstorm is getting stronger. Loud rain starts to pound the ground (write "2" on the board). Now we can hear the rumbling sound of thunder (write "4" on the board). The rainstorm is all around us! It is very loud! (wait a few seconds to enjoy the sound effects). The rainstorm is passing. The thunder stops (erase "4"). The loud rain stops (erase "2"). The wind stops (erase "3"). Even the soft rain stops (erase "1"). How quiet it is now that the rainstorm is over!

Using Tools: Choose a day that is breezy, but not too windy, for this activity. Before beginning, mark paper plates with N, E, S, and W, or help students do this. Review the concept of directions. Help students assemble their wind vanes, particularly the step in which the bare end of the feather is attached to the pencil eraser with the straight pin. Be sure the feather swings loose. Use damp sand if needed to hold pencils in place. When placing wind vanes outside, use a compass to locate north. For Think About It, students should feel the breeze hitting their faces if they face the opposite direction the wind vane indicates wind is blowing.

Learning Link: This activity explores climate and three geographic regions dominated by weather patterns: deserts, the arctic, and rainforests. Provide research materials about each region, and help students prepare and perform their plays.

Quick Lab Experiment

Weather This Week

PREDICT

What will the weather be like this week?

LAB TIME

1. Watch the weather each day for one week.

2. Show each day's weather. Cut and paste weather pictures from this page.

Some days, you will need to paste more than one weather picture.

THINK ABOUT IT

How did you dress for each day's weather?
Tell or write about it.

Weather This Week

Fill out this page as part of your lab.

LAB TIME

Show each day's weather. and the weather pictures from your lab page.

Monday	Tuesday	Wednesday	Thursday	Friday

THINK ABOUT IT

What is a season?

How might your lab answers be different if the season were different? Tell or about it.

Class Rainstorm

PREDICT

How can you make sounds like a rainstorm?

LAB TIME

Make the sounds of a rainstorm as your teacher tells a story!

1. Set up groups numbered 1, 2, 3, 4. Your teacher can help you.
2.  at the picture for your group. This is the sound you will make for the rainstorm. Practice making the sound.

1
soft rain
tap fingers on desk

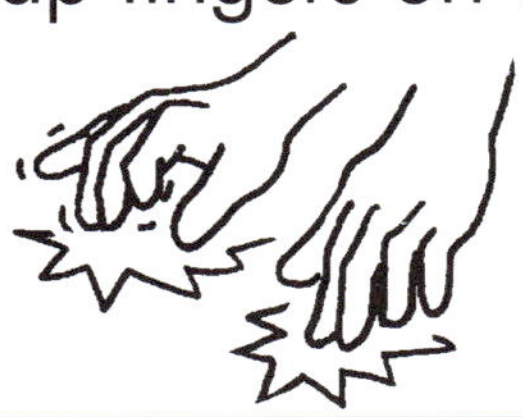

2
loud rain
clap hands

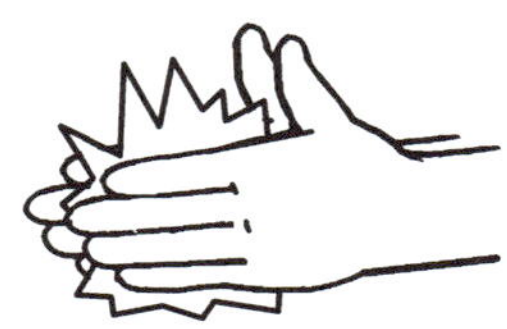

3
wind
blow air through lips

4
thunder
stamp feet
on floor

3. Your teacher will read a story about a rainstorm. When you see your number, act out your sound! Keep up the noise until your number goes away.

THINK ABOUT IT

Record a rainstorm with a tape recorder.
Tell or  about what you hear.

Homemade Wind Vane

Make and use a wind vane to find out wind direction.

The wind does not always blow in the same direction.
You can make a tool called a wind vane to find out wind direction.

LAB TIME

You'll Need:

flower pot　sand　paper plate　pencil with eraser　straight pin　feather

1.  the letters N, E, S, W for directions on your 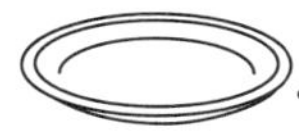. Your teacher will help you.

2. Fill the  with . Push the tip of the  into the .

3. Have your teacher push the  through the end of the  and into the eraser.

4. Go outside. Place the  on the 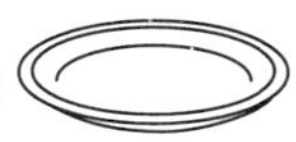 with the letters showing. Your teacher will help you.

5. When the wind blows,  at the . Which letter does the  point to? This is the wind direction!

THINK ABOUT IT

Can you think of a way to feel the direction of the wind?
Tell or  about it.

Weather Around the World

Some areas of the world have the same kind of weather most of the time. In these areas, weather controls how the land looks. It controls which plants and animals can live there. It controls how people live, too.

1. Learn about weather around the world. Read books or use the Internet. Your teacher can help you.

2. the name of each kind of place on the correct line.

RAINFOREST   ARCTIC   DESERT

It is hot and sunny here, without much rain.

____

It is cold and snowy here.

____

It rains almost every day here.

____

THINK ABOUT IT

Put on a play about the desert, arctic or rainforest. Dress the way you would for weather in your land.

The Sun

The sun, moon, stars, clouds, birds, and airplanes all have properties, locations, and movements that can be observed and described... The sun provides the light and heat necessary to maintain the temperature of the earth... Objects in the sky have patterns of movement. The sun, for example, appears to move across the sky in the same way every day, but its path changes slowly over the seasons. *National Science Education Standards*

Getting Started

To gather information online, enter the query: "solar system" AND "children" AND "learning," or "astronomy" AND "children." Also share background information below and books such as **The Sun, Our Nearest Star** (Franklyn Branley, 1988), or **What's Inside the Sun?** (Jane Kelly Kosek, 1999). Check the Equipment Table for materials needed to complete the activities in this unit.

About The Sun

- The sun is a medium-size star: an extremely hot ball of hydrogen and helium gas. Scientists estimate the sun to be about 4.5 billion years old.
- The sun is located in the center of our solar system. Objects that orbit the sun include nine planets and their moons, comets, asteroids, and other space objects. The sun is the closest star to Earth (150 million km or 93 million miles).
- The sun sends waves of energy, or radiation, to the earth; some of this radiation is released back into space, and some is trapped by the atmosphere and gravity. Light from the sun takes about eight minutes to reach our planet.
- As Earth rotates eastward each day, the sun appears to travel from east to the west through our sky. The sun "travels" the same distance each hour of the day. It is directly overhead at noon.
- The sun's energy helps create weather, seasons, and climates. It also makes life on Earth possible. Plants (which support all higher life forms through food chains) use the sun's energy to create their own food through photosynthesis.
- The sun's energy warms the earth and everything on it, but some things warm faster than others.

Misconception

Children may think the sun moves around the earth, when the reverse is true. In addition, they may not realize the sun never stops shining. We just cannot see it when it is night or when clouds cover it up.

Teaching The Unit

Quick Lab and Data Collection/Communication: Do this activity on a sunny day. Help students select sunny and shady spots to place their sets of objects for testing. Students should find that objects left in the sun feel warmer than those left in the shade. Results will vary based on what additional objects students choose to test, but generally the sun will heat metal objects the most and the clay the least. For Think About It, help students recognize that an object's exposure to sunlight and the materials from which it is made affect how much it is warmed by the sun.

Working Together: Do this activity on a sunny day. Start as early in the morning as possible, and continue for as many hours as possible. Help students select spots and draw arrows as reference points for later measurements. Help them draw lines to indicate the position and size of the shadow, and to mark the time the measurement was made. Students will see that shadows "move" as the day progresses. The sun's changing position in the sky causes this movement. When the sun is directly overhead (at noon), shadows are shortest. Explain that the sun is not really moving; the earth is rotating.

Using Tools: Help students read thermometers and practice using them to take temperatures. You might set up stations featuring each object for students to record temperatures, and allow students to choose a final object of their own to measure with the thermometer. Results will vary based on room conditions and the students' additional choices. For Think About It, students might try moving objects to change their temperatures.

Learning Link: For best results, use green, blue, red, or purple construction paper. Provide flat objects such as those listed for students to arrange on their paper. You may wish to have students tape them in place. Help students find a sunny spot out of the wind to place their art. Students will find that the sun bleaches the construction paper, and that the objects block the sunlight, resulting in dark shapes when they are lifted away.

Quick Lab Experiment

Sun Science

PREDICT

What happens when objects are placed in the  or away from the  ?

sun

LAB TIME

You'll need 2 of each:

bowl of water

black comb

metal spoon

aluminum foil ball

rock

clay

????
______
your choice

1. Place 1 set of objects in a spot where the  is shining.

2. Place 1 set of objects in a spot where there is shade.

3. Leave your objects for 1 hour. Feel each pair of objects.

THINK ABOUT IT

What is different about how the  shines on each set of objects?

Sun Science

Fill out this page as part of your lab.

LAB TIME

Which object feels warmer? Mark an "X" in the box to show your answer.

Object	In the Sun	In the Shade

???? ______		

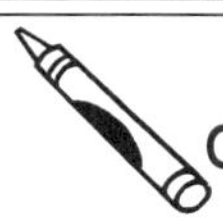 a ○ around the object that felt the warmest.  a ☐ around the object that felt the coolest.

     ???? ______

THINK ABOUT IT

Can you name 2 things that control how much the  heats an object?

Tell or  about it.

Shadows on the Move

PREDICT

How does the make shadows as it moves across the sky?

LAB TIME

Work with a partner.

1. Find a spot outside where the is shining.
2. Have your stand in the . an arrow on the ground with chalk. The should point the way your is facing.
3. for a shadow next to your .
4. Lay a meter stick where you see the shadow. Trace the shadow with . Write a number to show the hour of the day.
5. Every hour, stand on the . at the shadow. Trace the shadow using the and . a number for the hour next to the shadow. Do this as many hours as you can.

THINK ABOUT IT

What happened to the shadow from hour to hour?
What made the shadow change over time?
Tell or about it.

Thermometer

Use a to measure temperatures in your classroom.
thermometer

A is a tool that measures how hot or how cool something is. This measurement is called temperature. Numbers on a show temperature. The liquid inside a rises as temperature rises. It falls as temperature falls.

LAB TIME

1. Practice using a . Your teacher can help you.
2. Measure temperatures around your classroom. Place the in each spot for 3 minutes. Record the temperatures.

1, 2, 3, 4, 5 to order your objects from coolest to hottest.

Object	Temperature	Coolest to Hottest
glass of water	°	
window sill	°	
inside shoe	°	
inside desk	°	
your choice ________ ????	°	

THINK ABOUT IT

Can you think of a way to make your coolest object warmer?
Can you think of a way to make your warmest object cooler?
Tell or about it.

Sun Art

If you leave something in the , it heats up. Sometimes, things left in the also change color. Use this fact to make art!

You will need colored . You will also need flat objects to

paper

lay on your . Here are some objects you might try:

key puzzle piece coin button leaf

1. Pick a day when the is shining. Place your on the ground in the .
2. Place a on each corner of your to hold it down.

rock

3. Lay your objects flat on the . Order them any way you like.
4. Leave your in the for 3 or more hours.
5. Lift your objects off the paper. at the shapes they leave behind!

THINK ABOUT IT

at the parts of your that the sun could reach.
at the parts of your that were covered up.
How and why do these parts look different?
Tell or about it.

The Night Sky

Objects in the sky have patterns of movement... The moon moves across the sky on a daily basis much like the sun. The observable shape of the moon changes from day to day in a cycle that lasts about a month.

National Science Education Standards

Getting Started

To gather background information online, enter the search term "children" AND any of the following: "astronomy," "night sky" AND "learning," "moon," or "constellations." Also share books such as **Does the Moon Change Shape?** (Meish Goldish, 1989), **What Makes Day And Night**, (Franklyn M. Branley, 1986), or **Constellations** (Dennis Brindell Fradin, 1998). Share background information below as well. Check the Equipment Table for materials needed to complete the activities in this unit.

About The Night Sky

- Night occurs when our part of the earth turns away from the sun. We are in the earth's shadow. The earth blocks the sun. People on the opposite side of Earth experience daytime.
- Familiar night sky objects include the moon, stars, planets, clouds, artificial satellites, airplanes, and night-flying creatures.
- Our moon is called a satellite and is about one-fourth the size of the earth. The moon has no atmosphere and no water; life does not exist there.
- The moon revolves around the earth in 30 days. During this cycle, the moon undergoes different phases. It begins as a new moon and waxes into a crescent moon, half moon, and full moon during the first half of the month. It wanes into a half moon, crescent moon, and new moon during the second half of the month.
- The earth rotates on its axis, taking 24 hours to complete one rotation. As a result, the moon is only visible to us for a portion of each day. The moon rotates at the same rate it revolves, so we always see the same side.
- A star is a large ball of hot gas, thousands of millions of kilometers in diameter. Stars generate their own light. Almost all stars we see appear only as points of light, because they are so far from Earth.
- Constellations are patterns formed by bright stars that appear close to each other. Most constellations have names that were created by the ancient Greeks more than 2,000 years ago.

Misconception

Children may not know the moon does not generate its own light. Instead, it reflects light from the sun.

Teaching The Unit

Quick Lab and Data Collection/Communication: Inform parents of this activity and arrange to have them help conduct it on a clear evening. Prepare students by discussing differences between night and day. Be sure they understand the clues they are to look for in their scavenger hunt. Create a class chart to summarize and compare students' results. Allow time for students to discuss what people do after dark for Think About It.

Working Together: Before beginning this activity, tape a "day" picture (as described in the activity) on the wall on one side of the room. Tape a "night" picture on the opposite wall. Discuss what causes night and day on Earth. Have students form two groups, to represent the two halves of Earth. Students should stand in a circle facing outward. Help students decide which side of their pictures to show based on which wall picture they face. Have students rotate in place, remaining facing out, to model the earth's rotation. Remind them to change their pictures once they are facing a new picture on the wall. Discuss the connection to the fact that as the earth faces the sun, it is daytime. As Earth turns away from the sun, it blocks the light and falls into shadow, which is night time.

Using Tools: This activity demonstrates how the sun's light makes stars difficult to see during the day time. Students may believe stars rise and fall each night and day. In fact, they do not. Begin the activity with a brightly lit room. Gradually darken the room by closing shades, then turning off lights. Students should conclude that as the sun goes down and the sky darkens, the shining stars become easier to see. For Think About It, help students conclude big city lights can also make stars more difficult to see at night.

Learning Link: Begin this activity by sharing books and myths about constellations (define this term). Discuss the constellations shown and how their shapes gave rise to their names. Allow students time to share their star patterns and names with the class, as well as stories and songs they may know.

Quick Lab Experiment

Night Time Scavenger Hunt

PREDICT

What happens at night around your home?

LAB TIME

Go on a night time scavenger hunt! Have an adult go with you.

Can you find these things?

- 1 thing you can hear in the night air
- 1 thing you can smell in the night air
- 2 things floating or flying through the night sky
- 1 thing shining in the night sky
- 3 things that make night different from day

THINK ABOUT IT

Why is it dark at night?
Tell or write about it.

Night Time Scavenger Hunt

Fill out this page as part of your lab.

LAB TIME

Draw or what you found on your night time scavenger hunt.

What I found:

1 thing I could :

1 thing I could :

2 flying or floating things:

1 shining thing:

3 ways night is not like day:

THINK ABOUT IT

What are some things people do at night time?

or about what you think.

Night Time, Day Time

PREDICT

How does night turn into day?

LAB TIME

1. Make a night picture. a moon and stars on a piece of black paper with chalk.

2. Make a day picture. a sun and clouds on a piece of white.

3. Glue the back of the day picture to the back of the night picture.

4. Stand in a circle to be Earth. Your teacher can help you.

   Look at the picture on the wall. Hold up your matching picture.

5. With your group, turn in a circle like Earth. When the picture on the wall changes, change your picture.

THINK ABOUT IT

Finish this sentence: Night turns into day as ______________________

____

What clues in the sky tell you night or day is coming? Tell or about it.

Flashlight

Use a to show what happens to during the day.
flashlight

In the day, we cannot see . Where do the go?
They do not go anywhere! What happens? Use a to find out.

LAB TIME

1. Work with a . One holds a . One watches the .
   partner

2. Your teacher will make the room bright. Now it is day. Turn on the . The is like a . at the . Is its light easy or hard to see?

3. Your teacher will cover the windows. It is turning into night. at the . Is its light easy or hard to see?

4. Your teacher will turn off the lights. Now it is night. at the . Is its light easy or hard to see?

THINK ABOUT IT

Finish this sentence:

We cannot see in the day because

____

____

Your Own Stars

Did you know many groups of have special names? The names stand for how the groups look. Here are some groups and their names. Can you see why each got its name?

Hercules (man with club)

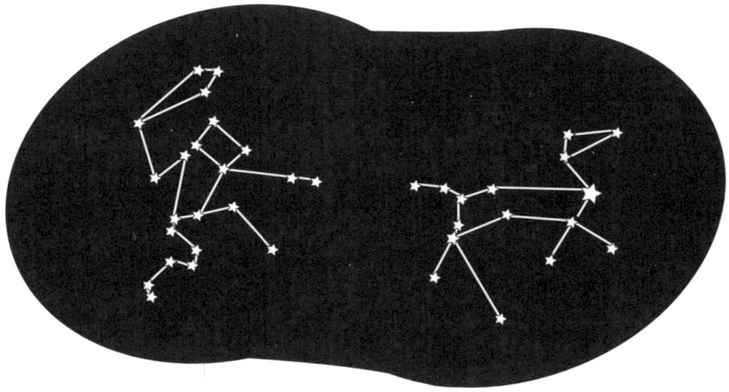

Canis Major (big dog)

Make your own group and name it!

You'll Need:

black paper pencil white chalk tape

1. Think about what you want your group to look like.
2. a picture on your with . Make your picture out of dots. Each dot is a . at the pictures above to help you.
3. Use the sharp tip of a to poke holes in each dot.
4. Use to the name of your group on the .
5. your picture to a window. at how the light shines through each !

THINK ABOUT IT

Do you know any songs or stories about stars? Tell, , or about them.